Figure 2 Filament extruder self-replicating rapid prototype RepRap done of Stainless Steel (SS) tube (l_{SS} = 1 cm) wiper motor (15 € https://www.ebay.com), motor speed switch controller Pulse Width Modulator (PWM) (9 < ΔV < 60 V, I_{DC} = 20 A https://www.ebay.com/itm/1pc-DC-9V-60V-20A-Pulse-Width-Modulator-PWM-Speed-Switch-Controller-For-DC-Motor/184532207984?hash=item2af6fa0970:g:mxoAAOSw2gBfqOAZ), power supply, auger (ø = 1.6 cm, l = 46 cm), thermostat (6 € https://www.banggood.com), relay (3.7 € https://www.ebay.com/itm/SSR-25DA-Solid-State-Relay-Module-25A-250V-DC-3-32V-Input-AC-24-380V-Output/163858866638?hash=item2626bfe1ce:g:7iYAAOSwGe5degT0), K-thermocouple (9.7 $ https://www.ebay.com/c/2254811970), metal screw fixing flexible mica-insulated band heater (P = 200 W, ΔV = 220 V, 3 cm x 3 cm, 4.5 $, https://www.ebay.com/itm/220V-30mm-x-30mm-Metal-Screw-Fixing-Flexible-Mica-Band-Heater-id/133580413800?hash=item1f1a035b68:g:1dUAAOSwsTVaojxE), two Aeolus fans (ø = 8 cm, ΔV = 12 V), axial ball thrust bearing, threaded rod (2x 10 mm), copper and silver tube (l_{Cu-Ag} = 1.5 cm), aluminium (Al) block chassis, low friction insulating PTFE guide, heat resistant Kapton tape, glass fibre cloth, filter net (ø = 200 µm), electroconductive hole (ø$_{Zn-Cu}$ = 1 mm), wood drill (l = 1 cm), wooden board for motor barrel and support (100 x 10 x 2 cm), three rocker switches, several screws and nuts, two sockets, two colours wires, inter-changeable nozzle, filament diameter 1.75 < ø < 3 mm, 0.6 g/mm ± 4.6 %, three-five-axis milling machine, v = 4.5 m/min, 0.5 < filament yield < 4.5 kg/h, ≈ 0.8 €/kg/h, 0.24 < specific energy < 1 kWh/kg making filament that cost < 1000 X than the bought, precisely, just 0.25 €/kg, 90 < price < 115 € - Image source: Xabbax and Ianmcill https://www.reprap.org/forum and, https://www.instructable.com

Recycled and Recycling Olympic 3D Bio-Printer (Recyclebot and RepRap-based)

© Copyright Antonio Silvestro, 2020

Recycled and Recycling Olympic 3D Bio-Printer (Recyclebot and RepRap-based)

© Copyright Antonio Silvestro, 2020

Recycled and Recycling Olympic 3D Bio-Printer (Recyclebot and RepRap-based)

Author:

Dr. *Antonio Silvestro*, self-employed heavenly based.
Department of *'Mathematic, Physics and Natural Sciences'*, Ba *'Biological Sciences'* University Federico II of Naples, Naples (NA), 80100, Italy.
Department of *'Agriculture'*, MSc *'Plant, Food Science and Environmental Biotechnology'*, University Federico II of Naples, Portici (NA), 80055, Italy.

Abstract:

Any recycled energetic discrete quantum, proportional in magnitude to the Revolutions Per Minute (RPM) of the *'Zeus - Genset (engine and generator)'* in motor mode (Kindle eBook 3.99 € Paperback 4.64 € https://www.amazon.com/dp/B08D3WJ1PF) flowing via the Hades infernal hot channels into nozzle as crystalline strings, cooled down by the Aeolus fan, solidified by the Boreas cold plate and coiled resting as Hermes bobbin borning or even rebirthing through Poseidon and Amphitrite vacuum pathways faster drying them, before bringing the shape of the designed CAD and desired physical body according the needs of any human consumer wishing for a quick, pragmatic and efficient tool for recycling any material by themselves at home just throwing rubbish in the Ares shredder and clicking on the icon of the lacking object on the Uranus display managed with Arduino IDE.

Keywords: engineering, biology, ecology, Arduino, 3D printing.

Correspondence for Copyright © permissions requirement to:
Dr. Antonio Silvestro born Friday 15th May 1992 at 20:00 under the Taurus sign ♉ ascendant Scorpio ♏ according the Greek specular to Monkey 猴 hóu rising dog (*Canis lupis*) Canis Major 狗 gǒu in Chinese and chestnut (*Castanea sativa*) in Druid, flower (*xochiti*) in Aztec astrology, North knot Capricorn ♑ goat (*Capra hircus*) 羊 yang and South knot Cancer ♋ rat (*Ractus norvegiensis*) 鼠 shǔ, resident in n°100 Nazario Sauro St., 80026, Casoria (NA) (Italia), number phone: +39 3382634244, emails: dr.antoniosilvestro@gmail.com, tonysilverxxx@gmail.com and antonio.silvestro5@studenti.unina.it.

Index

Recyclebot by now………………………………………………………………………………..2

Recyclebot – Uranus version………………………………………………………………….......6

- **Ares shredder**………………………………………………………………………..6
- **Transmission belt**………………………………………………………………….10
- **Hades heater**……………………………………………………………………….12

RepRap 3D Printer - Uranus version………………………………………………………….21

3D printing Computer Aided Design (CADs)……………………………………………….29

Recycled and Recycling Olympic 3D Bio-Printer (Recyclebot and RepRap-based)

© Copyright Antonio Silvestro, 2020

Recyclebot by now

The Uranus version of the Recyclebot for making recycled filaments for printing 3D physical objects with RepRap would have been characterized by a simple design with few pieces and low complexity economic, but high quality, thermoplastic filament versatility and reliability automated extruder.

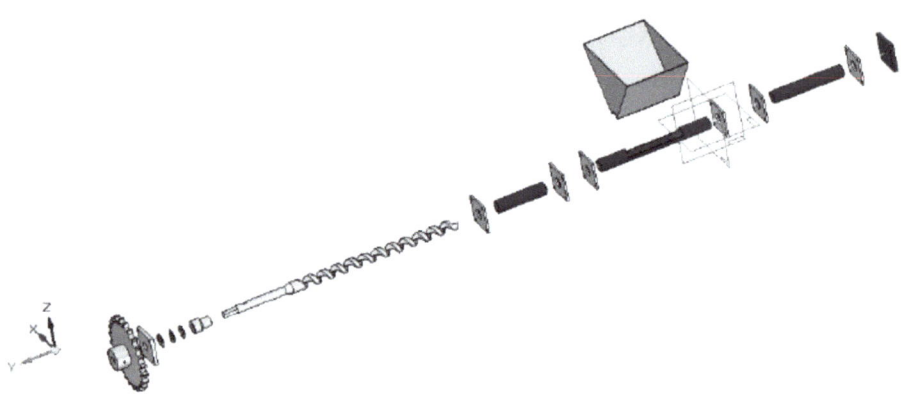

Figure 1 Recyclebot exploded assembly done of hopper, channel and screwing auger and gearing, large sprocket, and collar, (top). Recyclebot v5.0 (bottom) - Images source: https://www.appropedia.org/Recyclebot_v5.0

Recycled and Recycling Olympic 3D Bio-Printer (Recyclebot and RepRap-based)

© Copyright Antonio Silvestro, 2020

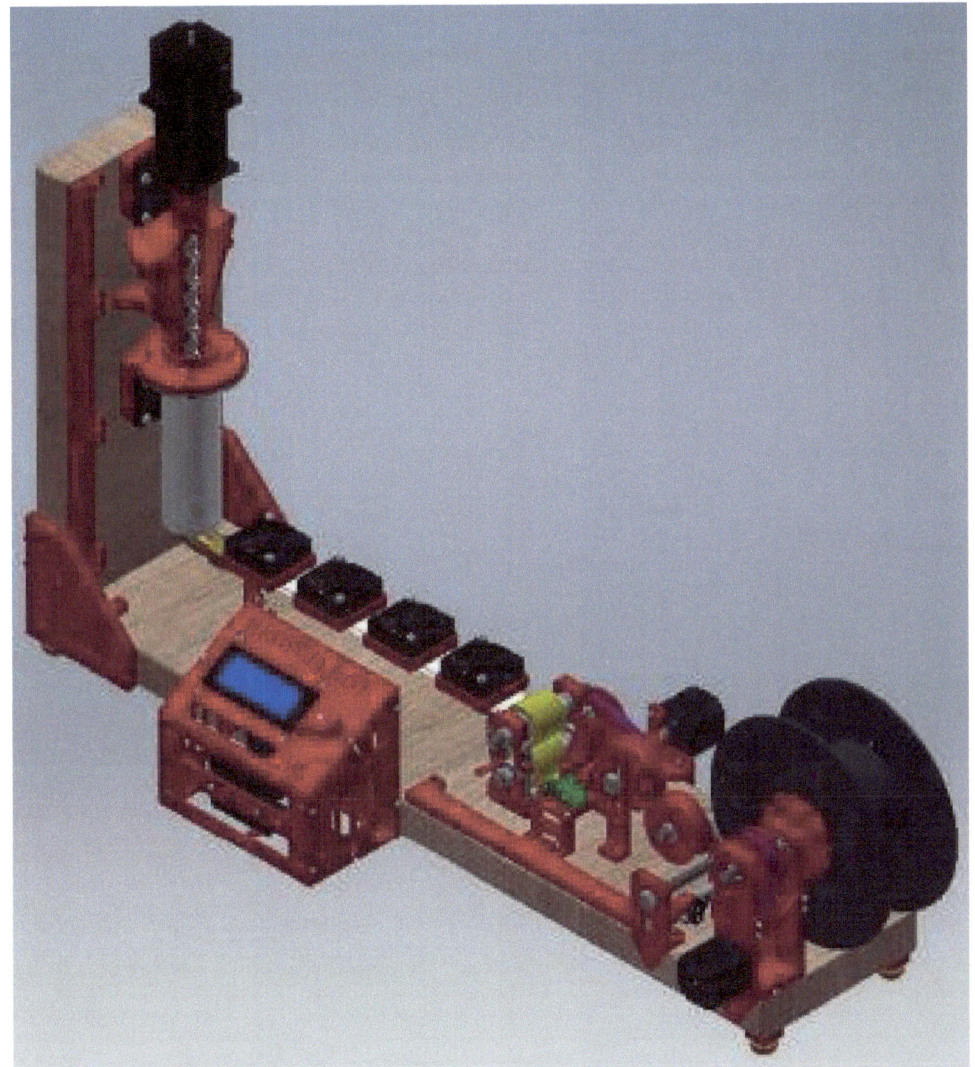

Figure 3 Recyclebot horizontal design (top), horizontal real (middle), vertical design (bottom) - Image source: *'RepRapable Recyclebot: open source 3-D printable extruder for converting plastic to 3-D printing filament'* by Aubrey L. Woern, Joseph R. McCaslin, Adam M. Pringle, and Joshua M. Pearce.

Recyclebot – Uranus version

Ares shredder

Ares would break-down, fragment and disaggregate any kind of hard matter living or not helping the natural recycling process of the colonized planet Earth, destruction-construction duality cycle in last instance homogenized by Poseidon. Fruits, seeds, bones, cellulosic paper, borosilicate glass, plastic insulators and conductive metals will not escape the sharp entangled series of Stainless Steel (SS)

Recycled and Recycling Olympic 3D Bio-Printer (Recyclebot and RepRap-based)

© Copyright Antonio Silvestro, 2020

circular saw blades (diameter = 40 cm, width = 3 cm; ⌀ : w = 10 : 0.75) of the God of war, 100 times powerful than the older heart of the father (P_{Ares} = 130 W). Common design can be shared by diverse applications differing for just few details like moving gears, depressing turbines/compressing impeller, pushing propellers, cutting blades, holding screws, transferring Archimedean screws conveyor, drilling auger. Blades and turbine/impeller, for example, differ for the fact that the first cut the fluid against which they move onto dividing it in two, while, the last just displace it, hence, they certainly will have two opposite tilts.

According the Hukki-Kick relationship between the physical corpuscles and the energetic work needed for breaking, those down the work of the *'Ares shredder'* would be equal to:

$$W = c\,(\ln\varnothing_i - \ln\varnothing_o)$$

Where:

$\varnothing_n$ = final diameter [m]

$\varnothing_o$ = originary diameter [m]

c = grinding coefficient

Archimedes of Syracuse famous for the hydrodynamic principle of displacement fair with volume of a physical body placed into water the screw that act as the super cells of Poseidon in the Nirvana Sharir of the man with siddhis capable to feel their lymphatic system recharge, the mirror that focalized the solar radiation for burning the ships as the planet Uranus would reflect the Sun frequencies onto Neptune at determined position among the elliptical orbits, not calculate in the present text, but guessed. Each real blade is in phase shit respect the following of the series for having a complex spiral structure, where the imaginary gap would bd filled by vacuum (*'Amphitrite-Poseidon – Vacuum pump and/or compressor'* © Copyright Antonio Silvestro, 2020 – Kindle eBook 0.89 € https://www.amazon.com/dp/B08D8H94K1) instead of regular air for accelerating the spinning velocity of the blades.

Minimal surfaces: plane, catenoid and helicoid. Logarithmic/exponential, elliptic, parabolic or hyperbolic function teeth blades comparative assay for evaluating to which one is associated the highest yield under the same conditions, variables according the state of matter that they have to collide with. Many small teeth would grind better than few would big, but the design would be durable just if they are enough resistant and durable, and without doubt, their design would be more laborious compared. 6-SS cogs of α = 30°, 45° and 60° (N teeth at variable angle than the x-axis).

Recycled and Recycling Olympic 3D Bio-Printer (Recyclebot and RepRap-based)

© Copyright Antonio Silvestro, 2020

	Logarithm	Exponential	Hyperbolic
Blade functions	$f(x) = \log_r x$, where: $r \in \mathbb{R}$	$f(x) = r^x$, where: $r \in \mathbb{R}$	$f(x) = \cosh x = (r^x + 1/r^x)/2$, where: $r \in \mathbb{R}$
Prime Derivative	$f(x)' = 1/f(x)$	$f(x)' = f(x)$	$f(x)' = \cosh x + \sinh x$
Incline angle	$\Theta = \arctan 1/(\ln(r))$	$\Theta = \arctan(r^x)$	$\Theta = \arctan(\cosh x + \sinh x)$
Tangent point between prime derivative and its function	$T = (x; \log_r x)$	$T = (x; r^x)$	$T = (x; [(r^x + 1/r^x)/2])$
Velocity in gas	$v_{(g)} = C_{rotor} \cdot RPM \cdot 0.0372 \cdot Re = C_{rotor} \cdot RPM \cdot 0.0372 \cdot \rho v x / \eta$ [mph]	Physical quantities	Density [g/cm³]
Air viscosity	$\eta_{air} = 0.0181$ mPa·s	Paper density	1.2 g/cm³
Velocity in liquid	$v_{(l)} = C_{rotor} \cdot RPM \cdot 0.0372 \cdot Re = C_{rotor} \cdot RPM \cdot 0.0372 \cdot \rho v x / \eta$ [mph]	Plastic	1.027 g/cm3
Water viscosity	$\eta_{water} = 0.89 \times$ mPa·s	Glass	2.5 g/cm³
Oil friction	$\eta_{oil} \approx 500$ mP·s	Metal	5 g/cm³
Velocity against solid	$v_{(s)} = C_{rotor} \cdot RPM \cdot 0.0372 \cdot Re = C_{rotor} \cdot RPM \cdot 0.0372 \cdot \rho v x / \eta$ [mph]		

Recycled and Recycling Olympic 3D Bio-Printer (Recyclebot and RepRap-based)

© Copyright Antonio Silvestro, 2020

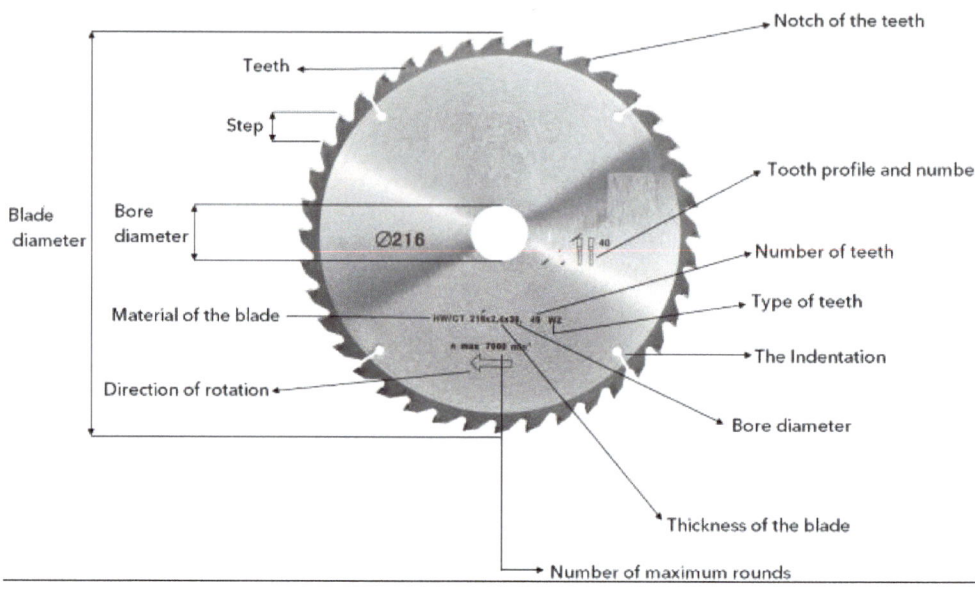

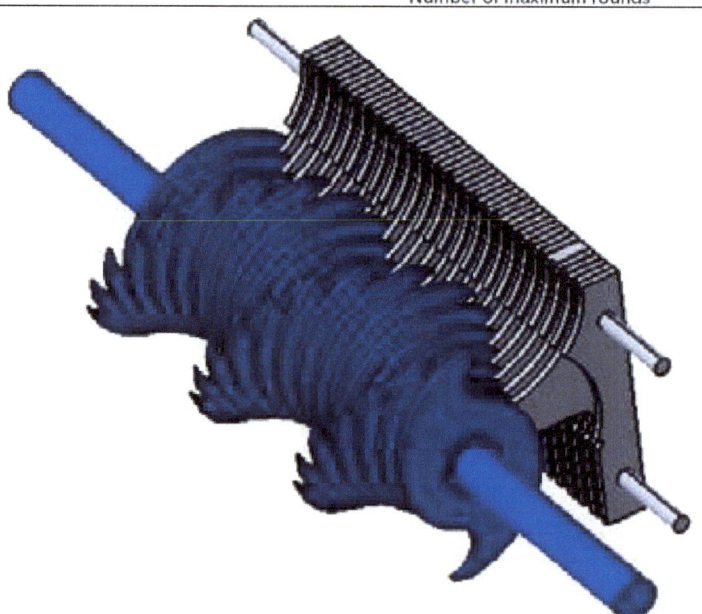

Figure 4 Ring blades (49/4.9 € https://italian.alibaba.com/product-detail/large-diameter-tungsten-carbide-saw-blade-cut-wood-shredder-blades-and-knives-62010194601.html?spm=a2700.galleryofferlist.normal_offer.d_image.5bf06b9bylg8z7) and spiral leadscrew done of spiral blades rotating within an internal cylinder [⌀ = 3 cm, length = 60 cm (shaft diameter: length = 1 : 20) https://www.wish.com/search/hexagonal%20stainless%20steel%20shaft/product/58cfe18e72573053380673db?source=search&position=28&share=web] – Images sources: https://powersawexpert.com/

Recycled and Recycling Olympic 3D Bio-Printer (Recyclebot and RepRap-based)

© Copyright Antonio Silvestro, 2020

Attrition and compression are the two main factors involved in the milling process being this last up to p = 300 MPa = 3 · 10^3 atm. A generic volume of matter would need 1 kW of power for being grinded properly (P : V = 1 kW : 1 L). The maximum noise value for this type of machine is SPL = 80 dB, higher than human voice, but certainly, lower than traffic jam.

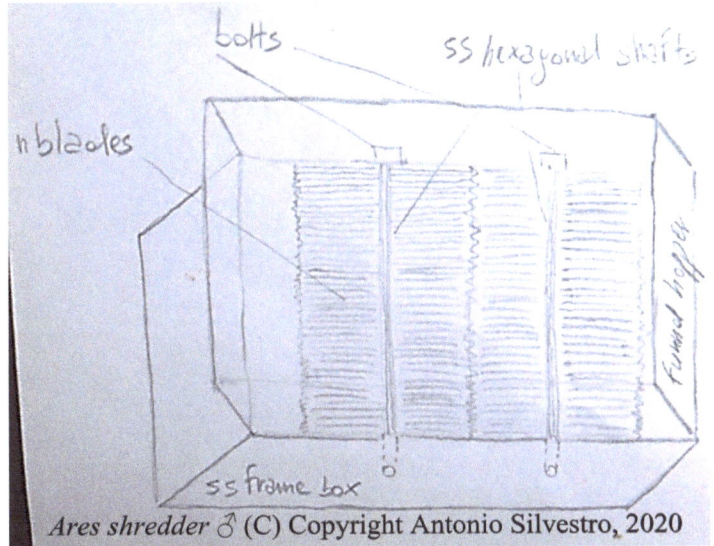

Ares shredder ♂ (C) Copyright Antonio Silvestro, 2020

Recycled and Recycling Olympic 3D Bio-Printer (Recyclebot and RepRap-based)

© Copyright Antonio Silvestro, 2020

Figure 5 *'Ares shredder'* would be designed by two parallel hexagonal shafts perpendicularly inserted in the two shorter side of the frame, holder by two thrusting bearing and bolted gears form one side and from a series of shredding blades, separated rings, to the other, more C-shaped supported to the inner side of the frame avoiding the improper physical object passage beside and not in between the fragmenting blades. The two polygonal (e.g. hexagonal) SS shafts (6 €
https://www.wish.com/search/hexagonal%20shaft/product/5901d6e8d0549419d774b716?source=search&position=5&share=web) avoid bolts or side blade triangular leaflets holders in the SS structural frame box keep maintaining them in place for properly spinning inside the funnelling hopper (7 €
https://www.wish.com/product/5cdd363895313f6634b3a9ba?hide_login_modal=true&from_ad=goog_shopping&_display_country_code=IT&_force_currency_code=EUR&pid=googleadwords_int&c=%7BcampaignId%7D&ad_cid=5cdd363895313f6634b3a9ba&ad_cc=IT&ad_lang=EN&ad_curr=EUR&ad_price=4.00&campaign_id=8705507239&retargeting=true&gclid=CjwKCAjwr7X4BRA4EiwAUXjbtzqaRfjMPBV6pZ5G60Htk7MkxYVEF0F6Vk-4OS-266uQ1kpbJE3oJRoCnukQAvD_BwE&share=web). N.B. Wolfram Carbide (WC) cutting tips (11 €
https://www.wish.com/search/Wolfram%20Carbide%20cutting%20tips/product/5ddddf2344bd8835241b6e8c?source=search&position=10&share=web) would be needed for making the holes in the box - Images sources: Jin Ling Wuan Guo (top) and *'Ares shredder'* sketch © Copyright Antonio Silvestro, 2020 (bottom).

Transmission belt

The gears connected to the *'Zeus Genset (motor mode)'* transmission belt for better regulating the environmental space occupied by the assortment 'Ares-Zeus' would be made according the following ratio for a functional *velocity*:

$$v_\varnothing = \frac{\varnothing_1}{\varnothing_2}$$

Where:

$\varnothing$ = diameter pulleys [m]

1 = bigger driving gear (wheel)

2 = smaller following gear (pinion)

The output *'Ares shredder'* speed (v_A), without any obstructive object, otherwise, maximal velocity, would be done by the ratio between the input *'Zeus motor'* velocity (v_Z) and the diametric ratio of the transmittive bears tightened by the chain, according the equation:

$$v_A = v_Z - v_\varnothing$$

Recycled and Recycling Olympic 3D Bio-Printer (Recyclebot and RepRap-based)

© Copyright Antonio Silvestro, 2020

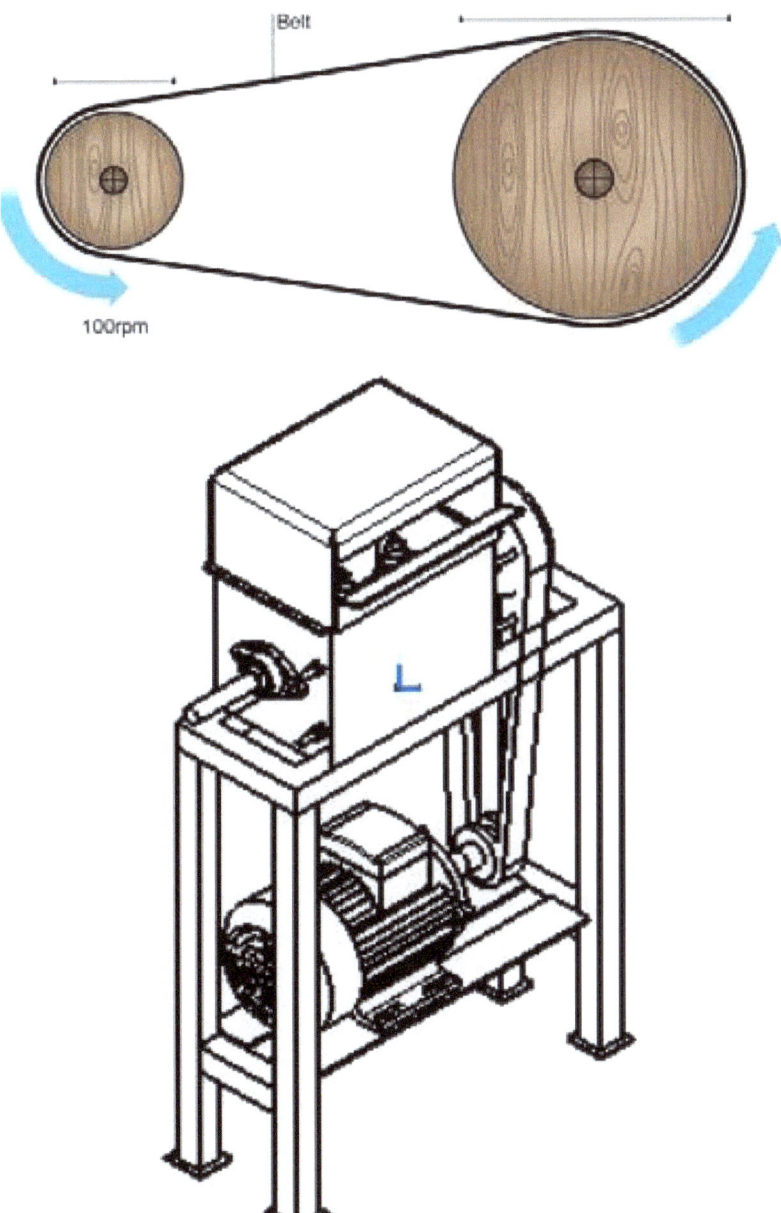

Figure 6 *'Ares shredder'* and *'Zeus motor'* sketch (top), V-belt driving transmission (7 €
https://www.wish.com/search/V-belt%20driving%20transmission/product/5dc3e85581771e115b6ae907?source=search&position=4&share=web) – Images source: DOI:10.1088/1757-899X/455/1/012119

Applications related to waste management like composting, faeces recycling or bones (Ca, P, Mg, Na, collagen, glycine) $4 < V_{frozen} < 12$ dm^3 = L, η = m/t = 2 tons/h, differentials filters with reducing

Recycled and Recycling Olympic 3D Bio-Printer (Recyclebot and RepRap-based)

© Copyright Antonio Silvestro, 2020

net diameters till the smallest of ⌀ = 0.22 µm. Ares parameters should be multiplied for 20 for grinding an adult male (*Ovis aries*) under the zodiac of Aries being its average volume $V_{ram} \approx 200$ L (Level P-1 : 2 km² < paper fragments < 5 mm² : P-7). Stainless steel (SS), macerator blades employed in this kit revolves at a high speed of RPM = 3600 min⁻¹ for reduction of human waste to a slurry one (85 € /7 L https://www.manomano.it/strumenti-per-raccolta-3866?model_id=2092522&g=1&referer_id=689772&gclid=CjwKCAjwjLD4BRAiEiwAg5NBFhu47ALAbTj70hNds_ClmpmCL4u_AaWuL65XaAdLVGSF05WJaXTKLBoCjl0QAvD_BwE).

Hades heater

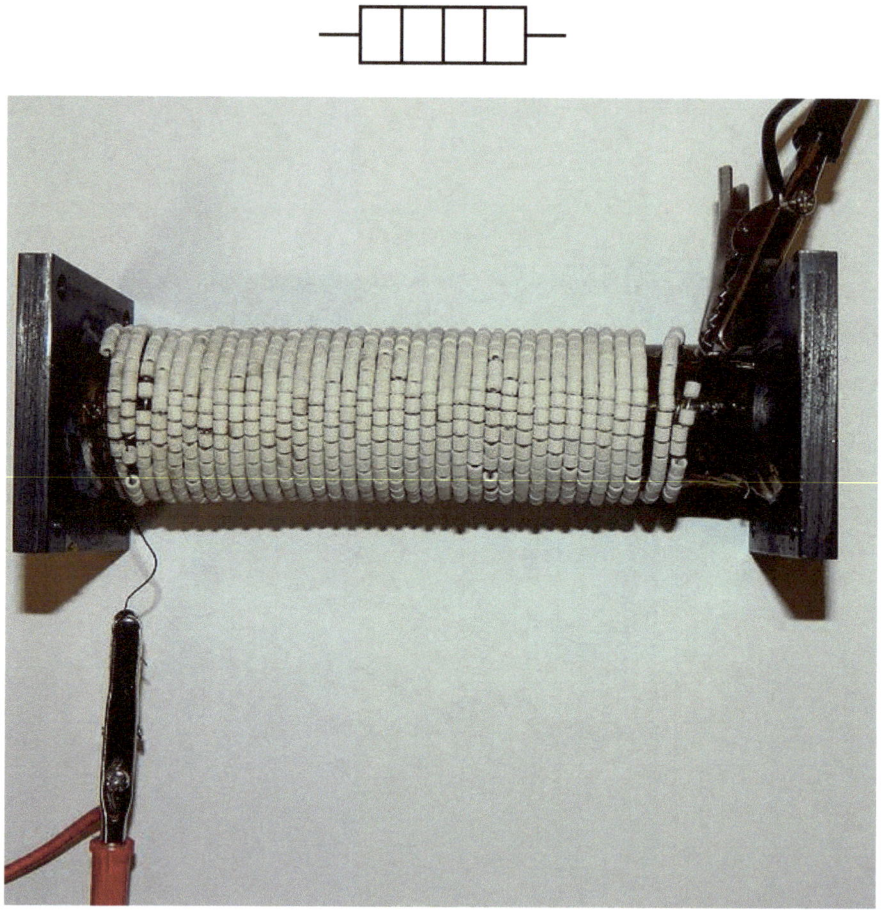

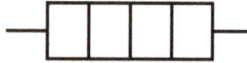

Figure 1 Heating elements pictogram (top). Hades thermo-conductive 14-gauge nichrome (Ni 80 %, Cr 20%), Kanthal (FeCrAl), cupronickel (CuNi), platinum (Pt), wolfram (W) or ceramic and semiconductors rings-insulated wire like Molybdenum DiSilicide ($MoSi_2$), Silicon Carbide (SiC), PTC silicon rubber, inductive coils wrapped around Persephone ferromagnetic aluminium (Al) or copper (Cu) barrel (l = 2 cm), secured with thermos-resistant brownish Kapton tape, transducing electricity into heat (0° < T < 80 ° C) undergoing resistive Joule Ohmic effect, supplied by AD/DC transformer (P = 75 W, I = 5 A, DC 12 < ΔV < AC 220 V, T_{out} = 225 °C) (bottom) – Image source: Anktivora, 2012.

Recycled and Recycling Olympic 3D Bio-Printer (Recyclebot and RepRap-based)

© Copyright Antonio Silvestro, 2020

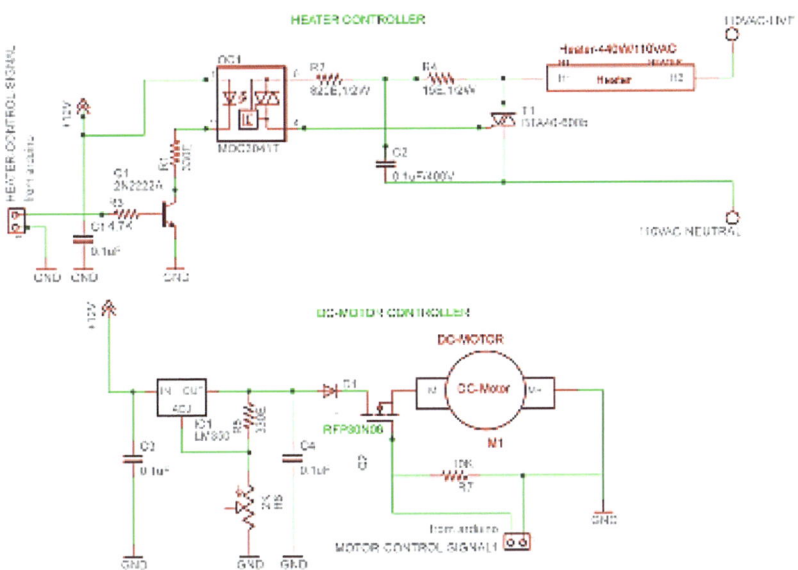

Figure 2 Pluto heater controller done of Caelus signal control connectors, triac, MDC204 socket (diode, switch and triac), two Saturn capacitors, three Venus resistor, three Gaia grounds, Vulcan heater *in sensu stricto*. Jupiter DC motor controller done of signal control connectors, voltage regulator, Caelus diode, two Saturn capacitors, Mars MOSFET, two Venus resistors and a trimpot, and four Gaia grounds – Image source: Ankitvora, 2012.

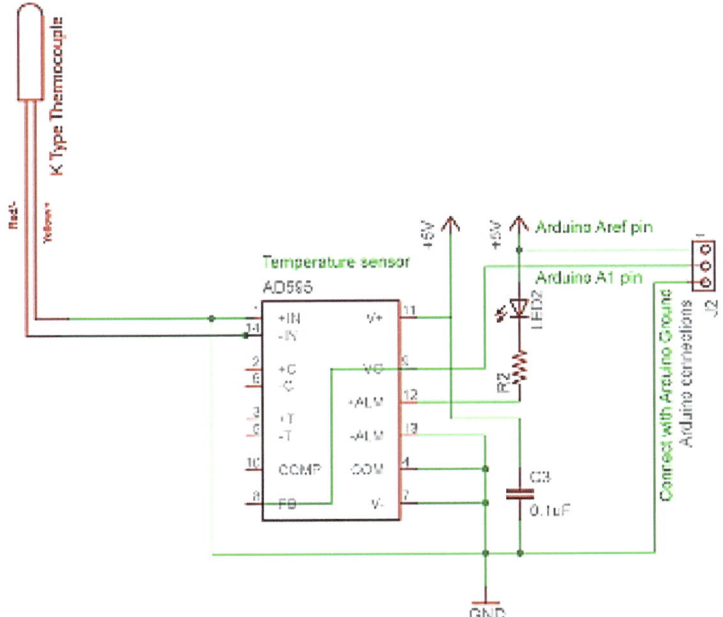

Figure 3 Venus temperature sensor AD595, K-thermocouple, LED, resistor, capacitor, Arduino connector – Image source: Ankitvora, 2012.

Recycled and Recycling Olympic 3D Bio-Printer (Recyclebot and RepRap-based)

© Copyright Antonio Silvestro, 2020

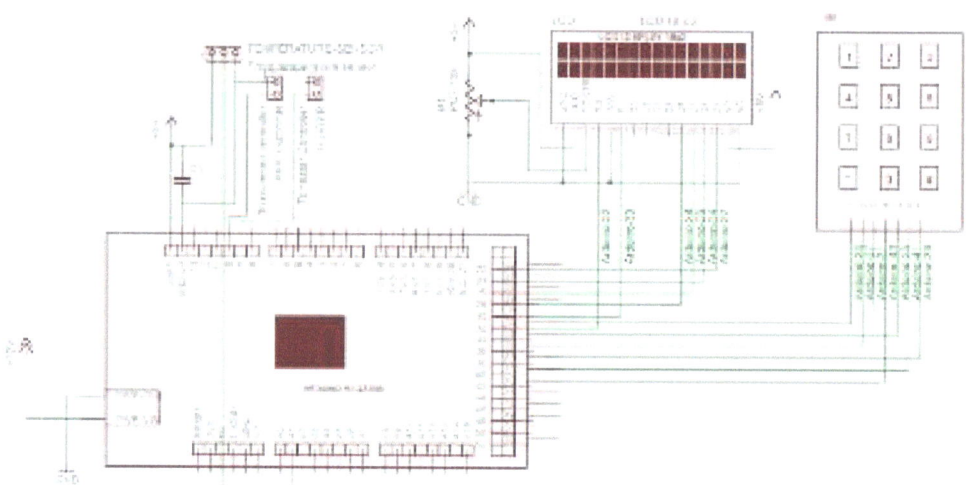

Figure 14 Uranus Arduino Mega 2560 controller, Aphrodite trimpot, LCD, keypad – Image source: Ankitvora, 2012.

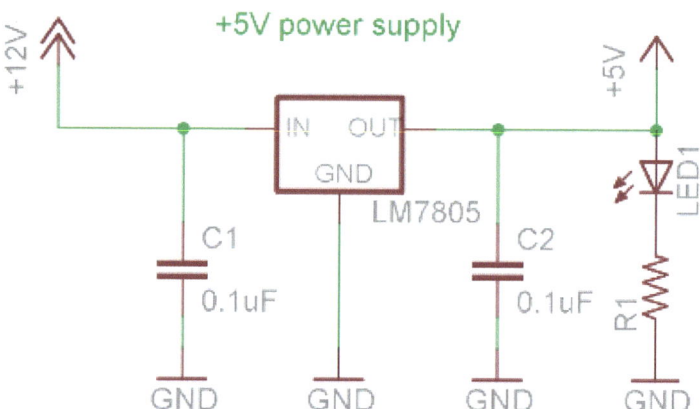

Figure 15 Neptune power supply circuit done of voltage regulator, LED, and a resistor two Saturn ceramic capacitor (P = 500 W, T = 500 °C) – Image source: Ankitvora, 2012.

Aeolus fan

If you imagine the Greek/Latin God of the wind Aeolus blowing into his right hand into the atmosphere, probably, you will see the resonating wave characterized by the follow frequency:

$$v_H = \frac{\omega_H}{2\pi} = \frac{\sqrt{\gamma \frac{P_0 A_c^2}{V_0 m}}}{2\pi} = \frac{v}{2\pi}\sqrt{\frac{A_c}{L_{eq} V_0}} = \frac{\sqrt{\gamma \frac{P_0}{\rho}}}{2\pi}\sqrt{\frac{A_c}{(0.3\emptyset + L_t) V_0}}$$

Recycled and Recycling Olympic 3D Bio-Printer (Recyclebot and RepRap-based)

© Copyright Antonio Silvestro, 2020

Where:

ω_H = angular frequency

v = speed of sound = 343 m/s

A_c = cross-sectional circular area

γ = Laplace coefficient, heat capacity ratio between the transitional state of matter, otherwise, adiabatic index: 1.4 for the troposphere (78% N_2, H_2, 21% O_2, 0.1% Ar, 0.03% CO_2) and its isolated constituent diatomic molecules.

p_0 = static pressure in the spherical cavity [atm]

V_0 = static volume in the spherical cavity [m^3]

m = mass [kg]

L_{eq} = length equivalent [m]

L_t = actual length of the tubular cavity [m]

⌀ = diameter [m]

The blades within the conical Ares shredder would break down the matter to recycle into small pieces, this, pushed through thermo-insulated Hades tubes triggered by the void created by the Amphitrite vacuum pump along the Poseidon auger spinned by the Zeus motor switched by the Arduino UNO ESP Wi-Fi, ensuring high reliability, efficiency with contained costs for producing recycled filament used by 3D Printer.

Recycled and Recycling Olympic 3D Bio-Printer (Recyclebot and RepRap-based)

© Copyright Antonio Silvestro, 2020

Part/Component	Link to supplier or source	Price [€]	Quantity	Comments
PCB				
Arduino Mega 2560 R3	https://www.ebay.com	10	1	
Arduino Uno clone with USB cable	https://www.wish.com	15	1	
SB400 Solderable PC breadboards	http://www.amazon.com/Solderable-BreadBoard-matches-tie-point-breadboards/dp/B0040Z3012/ref=pd_bxgy_e_img_y	7	2	
Screw terminals (l = 2.54 mm, pitch 2-Pin)	https://www.sparkfun.com/products/10571?	3.5	5	optional if you will be using breadboard or solder wires directly
Break away headers - straight	https://www.sparkfun.com/products/116	1.3	1	
Jumper wires pack (n = 140)	https://www.wish.com	5	1	optional, simple hook-up wires can also be used
Saturn ceramic capacitors (C = 0.1 uF, ΔV = 50 V)	http://www.digikey.com/product-detail/en/RPER71H104K2K1A03B/490-3811-ND/946447?cur=USD	1	4	
Saturn polyester/polyethylene film capacitor (C = 0.1 uF, ΔV = 630 V, DC radial type)	http://www.digikey.com/product-detail/en/PHE450MB6100JR06/399-5962-ND/2704616?cur=USD	1.2	1	
Uranus 1N4001 Micro Diodes (I = 1 A)	http://www.radioshack.com/product/index.jsp?productId=2036268	2.3	2	
Uranus LCD display (16 x 2 cm)	https://www.sparkfun.com/products/255?	12	1	
Keypad	https://www.sparkfun.com/products/8653?	3.5	1	
Neptune wall adapter power supply (ΔV = 9 V, I_{DC} = 650 mA)	https://www.sparkfun.com/products/298?	5.5	1	
Neptune discarded ATX power supply (P = 400 W)	https://www.ebay.com	15	1	for housing and DC motor power supply
Venus trimpot with Knob (T = 10 K)	https://www.sparkfun.com/products/9806?	1.7	2	

Recycled and Recycling Olympic 3D Bio-Printer (Recyclebot and RepRap-based)

© Copyright Antonio Silvestro, 2020

Voltage Regulator (ΔV = 5V)	https://www.sparkfun.com/products/107?	1	1	
LM350T - linear regulator	http://www.digikey.com/product-detail/en/LM350T/LM350TFS-ND/458688?cur=USD	0.8	1	
Mars transistor switching for power control N-Channel MOSFET (ΔV = 60 V, I = 30 A)	https://www.sparkfun.com/products/10213?	0.8	1	
Mars BJT transistors - NPN BC547	https://www.sparkfun.com/products/8928	0.7	1	
Uranus MOC3041M optoisolator Triac	http://www.digikey.com/product-detail/en/MOC3041M/MOC3041M-ND/281230?cur=USD	1	1	
Uranus Triac BTA40	http://www.digikey.com/product-detail/en/BTA40-800B/497-2405-5-ND/603430?cur=USD	9	1	
Venus resistor (R = 820 Ω, P = 1 W, 5 % metal oxide)	http://www.digikey.com/product-detail/en/ERG-1SJ821/P820W-1BK-ND/35939?cur=USD	0.3	1	
Venus resistor (R = 15 Ω, P = 1 W, 5 % metal oxide)	http://www.digikey.com/product-detail/en/ERG-1SJ150/P15W-1BK-ND/35755?cur=USD	0.3	1	
Venus resistor Kit (P = ¼ W, P_{tot} = 500 W)	https://www.wish.com	3	1	
Ares shredder				
Ring blades	https://italian.alibaba.com/product-detail/large-diameter-tungsten-carbide-saw-blade-cut-wood-shredder-blades-and-knives-62010194601.html?spm=a2700.galleryofferlist.normal_offer.d_image.5bf06b9bylg8z7	0.1	49	
SS shafts	https://www.wish.com/search/hexagonal%20shaft/product/5901d6e8d0549419d774b716?source=search&position=5&share=web	6	2	
	https://www.wish.com/product/5cdd363895313f6634b3a9ba?hide_login_modal=true&from_ad=goog_shopping&_display_country_code=IT&_force_curre			

Recycled and Recycling Olympic 3D Bio-Printer (Recyclebot and RepRap-based)

© Copyright Antonio Silvestro, 2020

Funnelling hopper	ncy_code=EUR&pid=googleadwords_int&c=%7BcampaignId%7D&ad_cid=5cdd363895313f6634b3a9ba&ad_cc=IT&ad_lang=EN&ad_curr=EUR&ad_price=4.00&campaign_id=8705507239&retargeting=true&gclid=CjwKCAjwr7X4BRA4EiwAUXjbtzqaRfjMPBV6pZ5G60Htk7MkxYVEF0F6Vk-4OS-266uQ1kpbJE3oJRoCnukQAvD_BwE&share=web	7	1	
Wolfram Carbide (WC) cutting tips	https://www.wish.com/search/Wolfram%20Carbide%20cutting%20tips/product/5ddddf2344bd8835241b6e8c?source=search&position=10&share=web	11	1	
Hades channel				
Furnace cement ($\Delta V = 100$ mL) and fireplace mortar	http://www.amazon.com/Meeco-Furnace-Cement-Fireplace-Mortar/dp/B000VZS0ZQ/ref=sr_1_1?	4	1	
Heat Sink	http://www.digikey.com/scripts/dksearch/dksus.dll?vendor=0&keywords=HS115-ND&cur=USD	0.9	3	
Red LEDs	https://www.sparkfun.com/products/533	0.6	2	
Thermocouple Type-K Glass Braid Insulated	https://www.banggood.com/K-Type-Thermocouple-Wire-Digital-Thermometer-Temperature-Sensor-Probe-Multimeter-p-1046076.html?utm_source=googleshopping&utm_medium=cpc_organic&gmcCountry=IT&utm_content=minha&utm_campaign=minha-it-en-pc&currency=EUR&cur_warehouse=CN&createTmp=1	5.8	1	
Thermocouple Amplifier AD595-AQ	https://www.sparkfun.com/products/306?	15	1	
High Temperature Ring Terminals	http://www.oemheaters.com/p-6032-high-temperature-ring-terminals.aspx	0.8	1	
Thermostat	https://www.banggood.com	6		

Recycled and Recycling Olympic 3D Bio-Printer (Recyclebot and RepRap-based)

© Copyright Antonio Silvestro, 2020

100 ceramic thermo-insulating beads (ø$_{int}$ = 0.4 mm, ø$_{ext}$ = 1.6 mm)	http://www.pescaloccasione.it	8.8	1
Nichrome wire	http://jacobs-online.biz/nichrome_wire.htm	5	1	approximately 11 feet required
Alligator Clips	https://www.sparkfun.com/products/111	1.8	4
Aeolus fan			
Aeolus cooling axial flow fan 30 < ø < 200 cm, p = 800 Pa = 8 matm			
Zeus Genset (motor mode) (Kindle eBook 3.9 €, Paperback 4.54 € https://www.amazon.com/gp/product/B08D3WJ1PF/ref=dbs_a_def_rwt_hsch_vapi_tkin_p2_i6)			
Wiper motor (ΔV = 12V)	http://www.banggood.com	29	1
Stepper motor (ΔV = 5 V)	http://www.banggood.com	5	1
Motor/Stepper/Servo Shield for Arduino kit	http://www.ebay.com	6	1
Irwin Tools 304300 93/4" x 17" Weld Tec Ship Auger	http://www.banggood.com	5.2	1
Koyo Torrington NTA-815 Needle Roller and Cage Thrust Assembly	http://www.amazon.com/Koyo-Torrington-Assembly-19000rpm-Rotational/dp/B006KT3084	5	1
Thrust Washer, TRA-815, .030-.032 X 1/2" I.D.	http://www.grainger.com/Grainger/INA-Thrust-Washer-4XFP9	3	2
Low Density PolyEthylene black pipe (LDPE 2 x 50 cm)	https://www.millstore.it	6.6	1
Cold metal sheet support (1.3 x 30 x 30 cm^2)	http://www.metalsdepot.com/products/hrsteel2.phtml?page=plate&LimAcc=$LimAcc	25	1
Roller Chain	http://www.thebigbearingstore.com/servlet/the-98/%2340-Roller-Chain/Detail	13	1
40B12 x 7/8" Finished bore sprocket 12 teeth #40 roller chain (pinion)	http://www.ebay.com/itm/40B12-x-7-8-Finished-bore-sprocket-12-teeth-40-roller-chain-/221109437613?#ht_500wt_1180	8	1
40B24 x 5/8" Finished bore sprocket 24 teeth #40 roller chain (wheel)	http://www.ebay.com/itm/40B24-x-5-8-Finished-bore-sprocket-24-teeth-40-roller-chain-/220952507390?	20	1
Scaffold			

Recycled and Recycling Olympic 3D Bio-Printer (Recyclebot and RepRap-based)

© Copyright Antonio Silvestro, 2020

Plywood bearing supporting plate	local hardware		as required	
Fasteners	local hardware		as required	
Nuts and bolts, screws	local hardware		as required	
Copper wire l_{Cu} = 65 cm	local hardware		as required	
Polyimide Kapton tape	http://www.wish.com	3	1	
		≈ 315		

RepRap 3D Printer - Uranus version

3D printing is an additive layer-by-layer manufacturing process for building three-dimensional objects giving born to artistic explicit or engineer parametric Computer-Aided Design (CAD) models, commonly, via the cheapest <u>Fused Deposition Modelling (FDM)</u>, invented by Scott Crump in 1989, for the extrusion through disposable nozzles moved by stepper or faster, but unnecessary servo motors of biothermoplastic filaments coiled, but also with the more expensive and less popular UV hardening of thermoplastic polymers, STereoLithography (STL), Vertically Integrated PRinted Electronics (VIPRE), metal high vacuum Electron Beam Melting (EBM), Liquid Additive Manufacturing (LAM), Selective Laser Sintering (SLS) and Melting (SLM) techniques. Wire Arc Additive Manufacturing (WAAM) is peculiarly suitable for metallic designs.

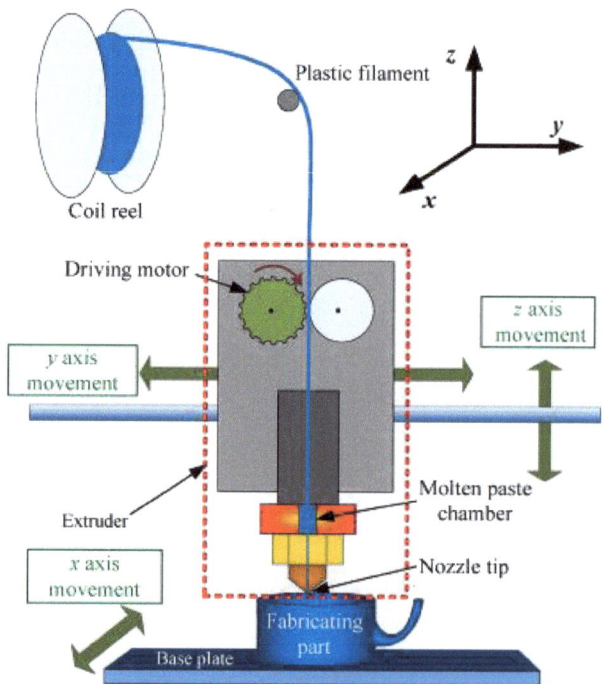

Recycled and Recycling Olympic 3D Bio-Printer (Recyclebot and RepRap-based)

© Copyright Antonio Silvestro, 2020

Figure 4 FDM diagram – Image source: https://doi.org/10.1016/j.addma.2015.10.001

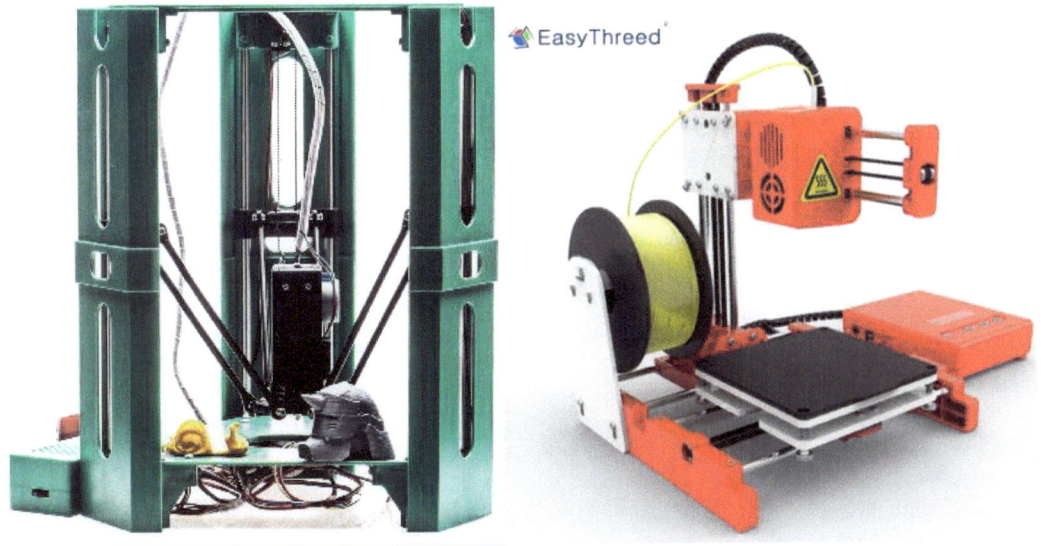

Figure 5 101 Hero Pylon 3D printer (three standing columns, top and bottom hexagonal cover versatile modules, screws, controller box, SD-TF card, extruder, 100-240 AC/12 DC transformer, PLA filament ø = 1.75 mm, tape and glass plate for 104 $ = 87.4 € - left) vs Easythreed X1 (25 x 21 x 21 cm, m = 1.3 kg, XZ axis, 10 x 10 x 10 cm print format, 180 < T < 230 °C, 10 < v < 40 mm/s, controller box, SD-TF card 512 Mb, 100-240 AC/12 DC transformer, P = 30 W, $ø_{nozzle}$ = 0.4 mm, PLA filament ø = 1.75 mm, holder, rod, cold square plate, input = STL/OBJ, output = GCODE, Operative System Windows or Apple Mac, USB cable, 70 € - right) - Image source: https://www.101hero.com/store/products/61949 and https://it.aliexpress.com/item/4001320662708.html?spm=a2g0o.cart.pcrcomd.7.52913c00G73eoF&gps-id=pcShopcartBuyagain&scm=1007.13440.197706.0&scm_id=1007.13440.197706.0&scm-url=1007.13440.197706.0&pvid=c6acffff-2675-4c2a-b128-d8ae04d45a21&_t=gps-id:pcShopcartBuyagain,scm-url:1007.13440.197706.0,pvid:c6acffff-2675-4c2a-b128-d8ae04d45a21,tpp_buckets:668%230%23131923%230_668%23808%234094%23597_668%23888%233325%236_668%232846%238109%23227_668%232717%237560%23273__668%233374%2315176%23331aliexpress.com/item/4000356622916.html?spm=a2g0o.store_home.productList_6000188401537.pic_4

An innovative FDM printers could be done of a Hermes plastic spool bobbin, two different sizes gears (wheel and pinon) around which the filament pass through entering in the aerographene (*'Electronic circuits fundamentals'* Kindle eBook 3.71 € https://www.amazon.com/dp/B086SFHDNV) insulated 'Zeus discharging chamber' where it become melted for finally dropping via the nozzles (0.3 < ø < 1 mm) onto Boreas cold plate for a quicker solidification.

The adhesion of the multiple layers and the time lasting of the model can be improved confining it into thermoplastic opaque or even metallic chamber filled with Modified Controlled Atmosphere (MCA) preventing photooxidation filled with inert nitrogen (N) or even nobles' gases.

Conical *nozzles* increase the flux speed decreasing the cross-sectional area of the empty channel in which wavy matter move along, from a coil placed in the far past to the future bright creation, the screening radius determine the properties of the strings jetted and printed. Certainly, a discontinuous

Recycled and Recycling Olympic 3D Bio-Printer (Recyclebot and RepRap-based)

© Copyright Antonio Silvestro, 2020

pulsing flow would be better for printing irregular objects, for which stepper motor become fundamental for being able to stop the extrusion where parts become weirdo, and would be useful to graph a torque (τ) dependant-thermocouple T(ΔV) Cartesian coordinate system. Nozzles in which UltraSounds (20 < v_{US} < 180 kHz) generated by piezoelectric transducers [(*'Dionysus/Bacchus – the spirit of enology and zymology'* Kindle eBook 0.93 € https://www.amazon.com/dp/B08F3V9SZD) and (*Neo wave – 'Fertile crescent'* (bead II) Kindle eBook 4.33 € https://www.amazon.com/dp/B08D87L14K)] can be used for spraying conventional 3D filaments, but also bio-inks, PCB flux, industrial varnishes, pharmaceuticals and Dye Solar Sensitive Cells (DSSC) coats (*'Wormhole in vitro'* Kindle eBook 7.74 € https://www.amazon.com/dp/B08DYDVZN9), fuels, catalyst into Proton Exchange Membranes (PEMs e.g. Nafion), Transparent Conductive Films [TCFs e.g. hard Indium/Fluoride-doped Tin Oxide (I/F-TO), flexible graphene, Ag nanowire, C nanotubes] (*'Illusion and reality about teleportation'* Kindle eBook 6.45 € https://www.amazon.com/dp/B08G53HLZN), and fluid culture mediums (*'Fluidoponics'* Kindle eBook 3.56 € https://www.amazon.com/dp/B08DL9B488).

Twist3D, based on the design of the widely extended RepRap BCN project's BCNozzle.

As the material slides inside the hot chamber 'Zeus Singularity' (e.g., T_{ABS} = 200 °C), its temperature increases, over glass transition value. At Stainless Steel (SS) removable nozzle with thin tip in which a membrane let flow filament produced fusing recycled matter with temperatures slightly lower than the melting temperature avoiding spraying liquid solutions. The heating multi parallel plates or even a unique helical barrier is creating by reducing the fins of the heat sink through which air pushed by the Aeolus fan circulates.

Temperature sensors *thermocouples*, done of insulating junctions between two or more different conductive thermopiles conductors generating a temperature/dependant voltage due to the electromagnetic field produced according T. J. Seebeck, assessing function via thermoelectric effect, are applied onto the Zeus liquefier for evaluate the degree of heat dissipation among the electroplates.

$$\Delta(T_{sense}) = \Delta V + \Delta V(T_{ref})$$

Insulated wires influence the way of changing of potential with the temperature, exempli gratia, Ni type E (Chronomel 90% Ni + 10% Cr – Costantan 55% Cu + 45% Ni) and Pt : Rh (40 : 20) alloy thermocouples are characterized for having the steeper and the smoother trend ΔV(T), respectively, the first capable of assessing ranges between - 270 < T < 1500 °C, while, the second between 0 < T < 2500 °C. - 185 < T < 300 < type continuous temperature range < 200 < B < 1700 °C. Using the following conversion formula these tools can be used for measuring pressure:

$$p = \frac{T_h (\Delta V^2 - \Delta V_0^2)}{\Delta V_0^2}$$

Where:

p = pressure [atm]

T_h = thermocouple constant

ΔV_0 = absolute temperature [V]

Recycled and Recycling Olympic 3D Bio-Printer (Recyclebot and RepRap-based)

© Copyright Antonio Silvestro, 2020

Recycled and Recycling Olympic 3D Bio-Printer (Recyclebot and RepRap-based)

© Copyright Antonio Silvestro, 2020

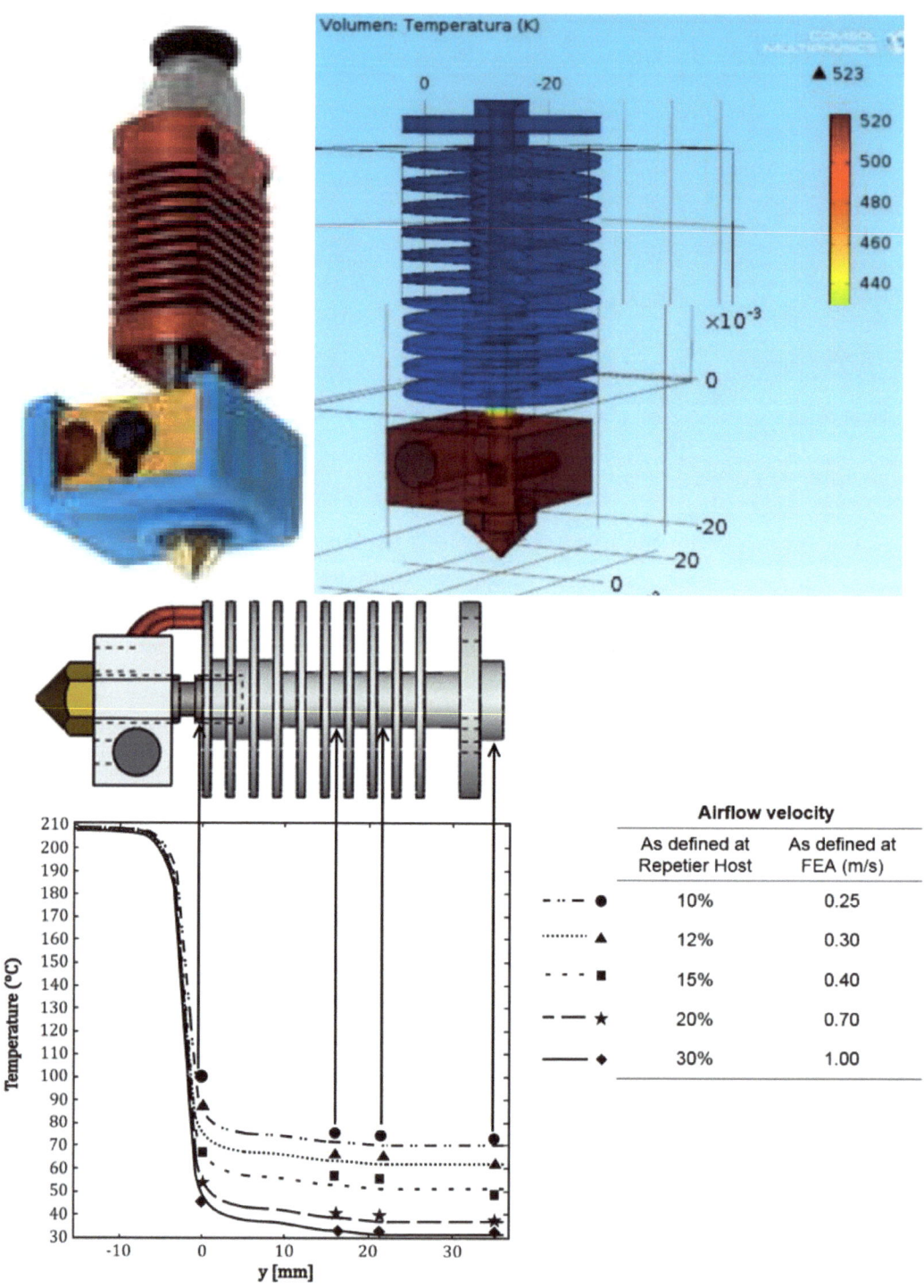

Recycled and Recycling Olympic 3D Bio-Printer (Recyclebot and RepRap-based)

© Copyright Antonio Silvestro, 2020

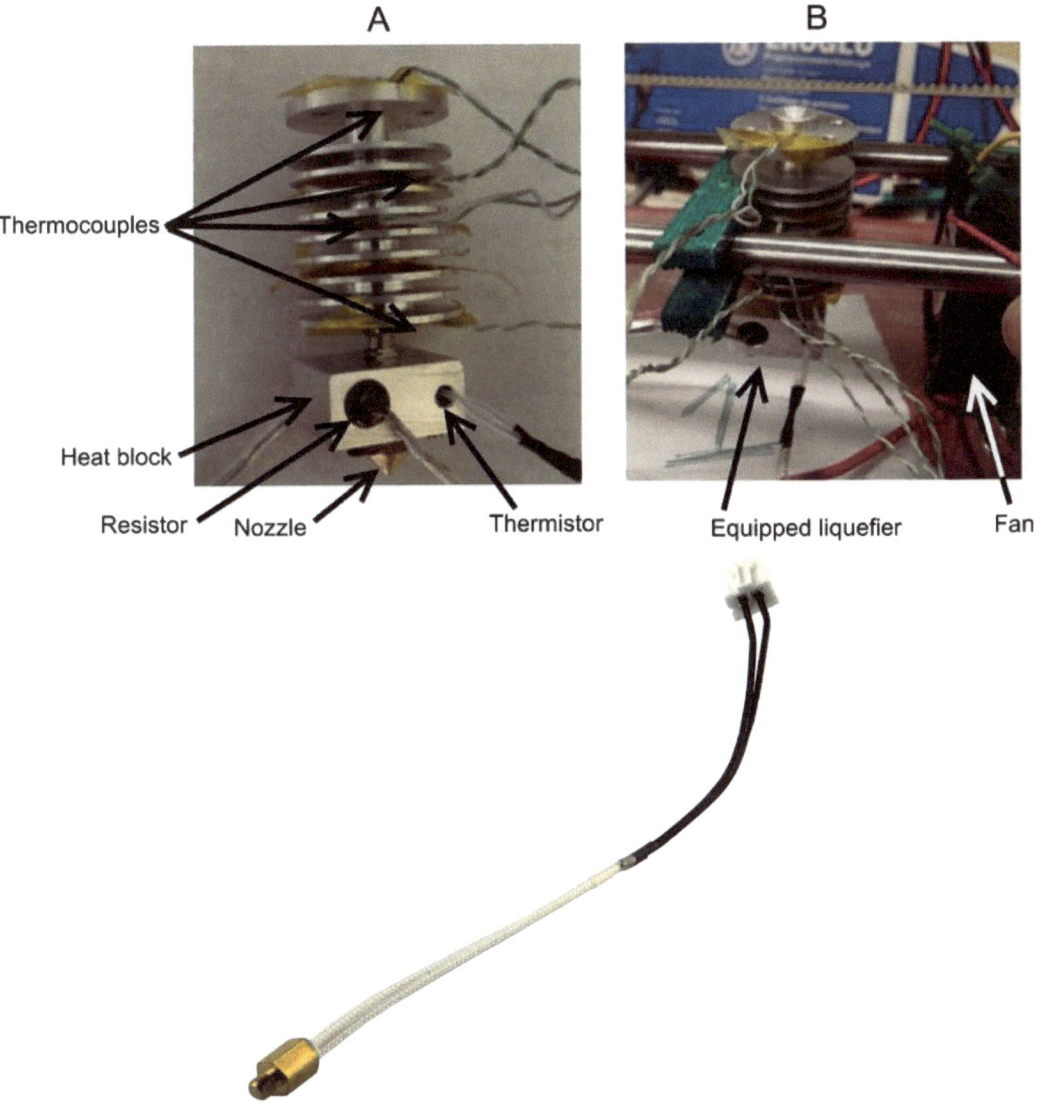

Figure 6 MK8 Hotend Extruder Kit (2 x 2 x 6 cm, $\varnothing_{filament}$ = 1.75 mm, $\varnothing_{nozzle}$ = 0.4 mm, thermocouples, heat block and nozzle for 2.35 €). Temperature diagram and airflow table. Extruder fan-cooled, with liquefier and highlighted main elements such as resistor and thermistor (0.96 €) – Images sources: https://it.aliexpress.com/item/4001133886678.html?src=google&albch=shopping&acnt=494-037-6276&isdl=y&slnk=&plac=&mtctp=&albbt=Gploogle_7_shopping&aff_atform=google&aff_short_key=UneMJZVf&&albagn=888888&albcp=1691306153&albag=64902423734&trgt=539263010115&crea=it4001133886678&netw=u&device=c&albpg=539263010115&albpd=it4001133886678&gclid=CjwKCAiA5IL-BRAzEiwA0lcWYqLoAdeToUrXwaS4cnIB0oCF6MFspam-DmF3fOkPRQ0ghfEiGhc1sRoCLQ8QAvD_BwE&gclsrc=aw.ds, DOI: 10.1016/j.mechatronics.2016.04.007 and https://www.dx.com/p/3d-printing-ntc-singleended-glasssealed-thermistor-temperature-sensing-line-with-terminal-2722889.html?tc=EUR&ta=IE#.X8Du_mhKjIU

Recycled and Recycling Olympic 3D Bio-Printer (Recyclebot and RepRap-based)

© Copyright Antonio Silvestro, 2020

Thermos-polymers needs to be extruded at glass temperature, when their amorphous shapes can become mesmerizing ordered crystalline physical object, but also metals and ceramics and even organic rubbers and lignin can be fused for shaping any object yearned. Multi-materials entities need to be specified in each 3D printing pixel element called 'voxel' for avoiding mixing the filaments, favouring homogenisation and the relative objects in any tiny cubic.

Aphrodite bio-inks made of liquid mixture of cells, matrix (hydrogel alginates, combination of acid with a UV- initiated PV-based cross-linker, nanocellulose, nutrients, oxygen and polymerizing calcium) are deposited layer by layer for creating biomaterials utilized in biomedical drug trials and included whole tissues and organs replacements. Injecting Bioprinting technique, patented by Thomas Boland in 2003 may remove the need for immunosuppressive drugs and decreases the probability of transplant rejection. However, since it may not always be possible to collect all the needed cell types, it may be necessary to isolate STEM cells or induce pluripotency in harvested tissues (2).

As the *Veneridae* seashells in the oceans of Poseidon, filtrate and purify the heavy metals remnants of the war of Ares felt on the bottom since the endless time of Chronus, before entering back in the energy-matter circle on the planet ruled by Gaia, passing through the tunnels of the Hades' underworld escaping the Hell as Orpheus and Eurydice for rebirthing as beautiful Aphrodite bio-filaments slowly melting into the Hephaestus nozzle cooled by the Aeolus fan drying-out the wet death. Now, in the space of your eyes you are going to read the methods for generating organic bioplastic that would shape into filaments by the Recyclebot and later used by the RepRap 3D printer:

1) Boil peel m = 25 g e.g., *Citrus sinensis*, *Cucumis lanatus* or *Coffea arabica* and $V_{Salvia\ officinalis}$ Oil = 5 mL, V_{H2O} = 15 mL, corn-starch flour $m_{Zea\ mays}$ = 50 g, m_{NaCO3} = 2 g, $V_{Citrus\ limoni}$ = 5 mL, $V_{CH3COOH}$ = 5 mL, V_{H2O} = 15 mL.

2) Boil peel m = 25 g e.g. *Musa banana* 3 x V_{becker} = 500 mL, $V_{cylinder}$ = 10 mL, funnel, glass rod, $V_{micropipette}$ = 1 mL, gloves, sodium meta-bisulfide ($NaHS_2$) extraction solvent, V_{H2O} = 375 mL, boil and collect at T = 30 °C, decant the water, backed in hoven within aluminium (Al) parched Petri dish for t = 30 min at T = 120 °C, blend the peels, V_{HCl} = 3 mL [HCl] = 0.5 M into V_{becker} = 140 mL, plasticizer $V_{glycerine}$ = 2 mL, V_{NaOH} = 3 ml [NaOH] = 0.5 M, dry in the hoven within aluminium (Al) parched dishes for t = 1 h at T = 130 °C (https://www.youtube.com/watch?v=cyfLoJBt670).

Fluoro-polymers porous bio-insulator [e.g., Ethyl Tetra Fluoro Ethylene (ETFE), Fluorinated Ethylene Propylene (FEP)]: fermenting methanogens microorganisms (e.g. *Thermotoga neapolitana*) releasing methane, then chloralhydrate, fluoro-hydrate, finally, polymerizing into bio-plastic.

White, thick and opalescent PTFE is made from methane (CH_4) in a series of reactions:

a) production of trichloroethane (chloroform)

b) production of chlorodifluoromethane

c) production of TetraFluoroEthene (TFE)

d) polymerization of TFE

Recycled and Recycling Olympic 3D Bio-Printer (Recyclebot and RepRap-based)

© Copyright Antonio Silvestro, 2020

Aphrodite filament thermoplastic materials	Glass temperature T_g [°C]
Tire rubber	-70
PolyVinyliDene Fluoride (PVDF)	-35
PolyPropylene (PP)	-10
PolyVinyl Fluoride (PVF)	-20
Poly-3-HydroxyButyrate (PHB)	15
PolyVinyl Acetate (PVA)	30
PolyChloroTriFluoroEthylene (PCTFE)	45
PolyAmide (PA)	47
PolyLactic Acid (PLA)	60
PolyEthylene Terephthalate (PET)	70
PolyVinyl Chloride (PVC)	80
PolyVinyl Alcohol (PVA)	85
PolyStyrene (PS)	95
Acrylonitrile Butadiene Styrene (ABS)	105
PolyMethylMethAcrylate (PMMA)	105
PolyTetraFluoroEthylene (PTFE) 'Teflon'	115
PolyCarbonate (PC)	145
PolySulfone	185
PolyNorbornene	215
Hight-Density PolyEthylene (HDPE)	-120
PolyUrethane (PU)	-63
PolyPhenylSulphonete (PPSU)	220
PolyEthylene Terephthalate (PETG)	85
High Impact PolyStyrene (HIPS)	100
PolyTetraFluoroEthylene (PTFE)	119
PolyEtherEther Ketone (PEEK)	140
PolyUrethane (PU)	60
PolyEthylene (PE)	-110

N.B. T_g is the temperature at which a polymer hardens themselves (amorphous -> crystalline). The values are obtained at sea level pressure (p = 1 atm). In green PP and PVF materials that easily can be toughened with fridge-freeze capacitors, and among them PP remain the most available and diffused on the market.

Recycled materials like Laywood can be used for making filament using the Fused Particle Fabrication (FPF), but it is possible to use biotechnological feedstock for isolating biopolymers extruded from organisms like *Tithonus Insecta* mealworms (*Tenebrio molitor*) (*'Culture and cultures'* KDP Amazon eBook 3.88 € https://www.amazon.com/dp/B08DL76M5T and/or Google Play Books Partner eBook 6.36 € https://play.google.com/store/books/details?id=jZ4lEAAAQBAJ).

RepRap 3D printer (270 x 205 x 210 mm^3, P = 60 W, I = 5A, ΔV = 12V, v = 15 mm^3/s.

Snappy v3.0 (ΔV = 12 V, P = 120 W, 5 stepper motors, cooling fan, extruder, borosilicate glass plate, 275 $).

Among the plenty applications of the 3D printing there are: toolroom, rapid prototyping, food (e.g. Spaghetti pasta processing. Clothing, shoes, robots, circuit boards, tissue bioengineering. The

Intellectual Property (IP) coming from the utilization of this technique can undergo legal iter like patents, industrial designed rights, copyrights and trademarks.

Typical layer resolution = 100 µm = 250 Dots Per Inch (DPI)

ISO/ASTM52900 - 15 defines seven categories of Additive Manufacturing (AM) processes within its meaning: binder jetting, directed energy deposition, material extrusion, material jetting, powder bed fusion, sheet lamination, and vat photo-polymerization.

Import free filament extruder CAD parts in FreeCAD, Copyright, STEP format

Draw the filament extruder model lacking mates

3D printer e-commerce print filament extruder

https://grabcad.com/library/tag/reprap and https://www.thingiverse.com/search?q=reprap&dwh=555df264a1e5c38

Import Printer free draws, sketch 3D printer lacking parts, let printer the 3D printer

Rendering image synthesis of 2-3D models in scene files containing geometries, viewpoints, texture, lighting and shading, suitable for video games, simulators and architectures designs.

3D Printer Creality, Tronxy brand Ender2, T819or EZT model (Al frame ≈ 40 x 40 x 60 cm, power P = 180W) - 5 < $T_{environment}$ < 40 °C, 20 < humidity < 80 %, has nozzle temperature: 170 ≤ T ≤ 275 °C, hot plate temperature: 00 < T < 110 ° ($T_{ABS\ nozzle}$ = 230 °C, $T_{ABS\ plate}$ = 80 °C: $T_{PLA\ nozzle}$ = 205 °C, $T_{PLA\ plate}$ = 50 °C), 1 < v < 20 cm^3/s, LCD screen, card reader, USB port connect and slicing software: Simplify3D, Cura, repetier-host, slic3r.

Uranus let felt on *Gaia* LUCA, which microalgae descendant would be used for 3D printing.

The first step of a 3D bioprinting is the model making, usually, with Computed Tomography (CT) and Magnetic Resonance Imaging (MRI), in a 3D printable file in the format. stl. then used for photo-/stereo-lithography, magnetic or direct cell extrusion bioprinting (Organovo&Aspect biosystem company - http://organovo.com; https://www.aspectbiosystems.com).

Recently, *'Scotty'*, a 3D teleporter has been invented in the Hasso Plattner Institute (Brandenburg – Germany), by the research team formed by Stefanie Mueller, Martin Fritzsche, Jan Kossmann, Maximilian Schneider, Jonathan Striebel and Patrick Baudisch. The machine consists of an off-the-shelf 3D printer, like a *'MakerBot Replicator 2X'*, that use a Raspberry Pi processor unit to which is added a 3-axis milling machine Arduino-controlled, a camera, and an encryption micro-controller. Phases of destruction, scanning and printing alternating themselves till the end of the transmission process (https://www.youtube.com/watch?time_continue=6&v=Qtp7kkKXMOw).

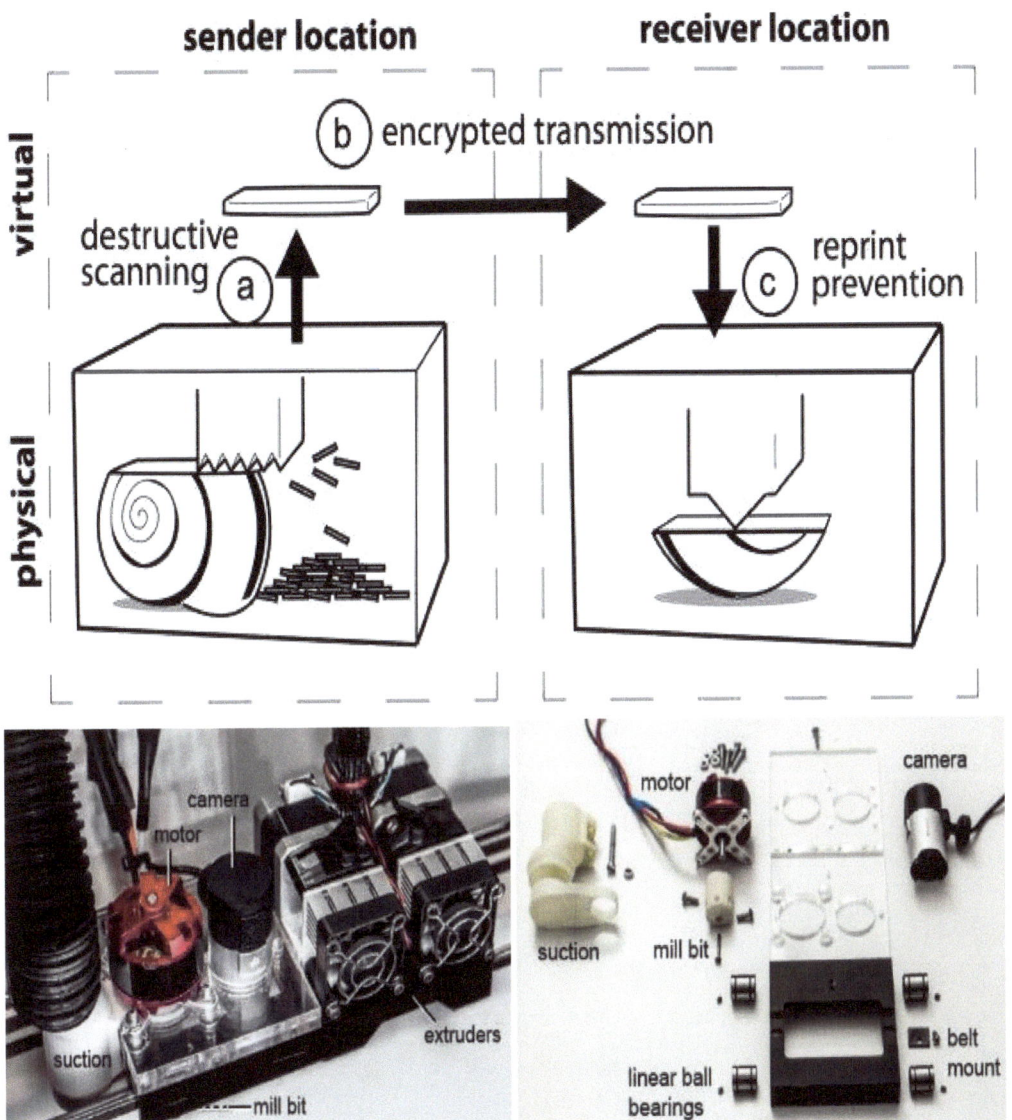

Figure 17 *'MakerBot Replicator 2X'* draw (top), *'Scotty'* device (bottom left), the Milling Apparatus (bottom right).

Teleporting delivery of any products aproximable to a physical object would become faster and ecologically friendly avoiding pollution due to transports, relatively easier would be the money teleportation as they are bi-dimensional and standardize digital model. Organs and tissue transplantation like the myocardium transfer could be made leaving the anesthetized patient, lying on the stretcher with the ribs cage open, while, the 3D printer drops bio-ink down directly in the long-suffering chest. Little nozzles acting in concert, each printing a patch of the extruded or injected organ

Recycled and Recycling Olympic 3D Bio-Printer (Recyclebot and RepRap-based)

© Copyright Antonio Silvestro, 2020

could increase the bioprinting efficiency and making the whole process more precise, even, for the stretchable cavities of variable volume vessels and valves.

3D printing Computer Aided Designs (CADs)

Commercial CADs: AgiliCity Modeller, Autodesk AutoCAD, Bricsys BricsCAD, Dassault Systems CATIA, Dassault Systems SolidWorks, Kubotek KeyCreator, PTC Creo (formerly known as Pro/ENGINEER), Siemens Solid Edge, Trimble SketchUp, Alibre Design, AllyCAD, Autodesk Inventor, AxSTREAM, Bentley Systems - MicroStation, Cobalt, IRONCAD, MEDUSA, Onshape, ProgeCAD, Promine, PunchCAD, Remo 3D, Rhinoceros 3D, RoutCad, Siemens NX, SketchUp, SpaceClaim, T-FLEX CAD, TurboCAD, VariCAD.

Freeware and open source: 123D, BRL-CAD, BricsCAD Shape, FreeCAD, LibreCAD, QCad, OpenSCAD, SolveSpace, Windows 3D Builder.

CAD kernels: Parasolid by Siemens, ACIS by Spatial, ShapeManager by Autodesk, Open CASCADE, C3D by C3D Labs, Cinema 4D, Maya, 3ds Max, Blender, LightWave, Modo.

Clothing specific: Marvelous Designer, CLO3D, Optitex.

3D printing **file sharing** platforms include: Shapeways, Sketchfab, Pinshape, Thingiverse, TurboSquid, CGTrader, Threeding, MyMini Factory, and GrabCAD.

Online **marketplaces** for 3D content allow individual artists to sell content that they have created, including Shapeways, TurboSquid, CGStudio, CreativeMarket, Sketchfab, CGTrader and Cults.

CAD models **groups**:

- <u>Parametric (Solid)</u>: cubic network that fits better with rigid, squared, firm models often by constructive assembly of primitives or more seldom via their subdivision, or even by implicit surfaces for medical and engineering;

- Shell Polygonal Mesh (Non-Uniform Rational B-Splines): manifold arrangement better for smooth, circular, curved shapes for games and films.

There are three main **file types** used in 3D printing:

1. IGS or IGES (Initial Graphics Exchange Specification) adopted by the American National Standards Institute (ANSI).

2. ASCII or Binary STL (STereoLithography or Standard Tessellation Language)

3. AP214 <u>STEP</u> (Standard for the Exchange of Product model data)

The originary STEP file would be converted in G-code files with a slicer software, each corresponding to a slice of the 3D model designed, and send to embedded MicroController Unit (MCU) on Metal Oxide Semiconductors (MOSs) of an monolithic Integrated Circuit (IC) chips characterized by e a Control Process Unit (CPU), Random Access Memory (RAM), Input/Output serial ports, oscillating clock generator (e.g. $V = 0.04$ mm^3, $P = 16$ nW wireless and batteryless Cortex-M0+ processor for opto-thermically measurements) that manage the motors.

Recycled and Recycling Olympic 3D Bio-Printer (Recyclebot and RepRap-based)

© Copyright Antonio Silvestro, 2020

Coming soon 3D printed physical objects suitable for electronics like Micro SD (512 GB/12 € https://www.wish.com/product/5d9fdc0df6c318513fde36bd?from_ad=goog_shopping&_display_country_code=IT&_force_currency_code=EUR&pid=googleadwords_int&c=%7BcampaignId%7D&ad_cid=5d9fdc0df6c318513fde36bd&ad_cc=IT&ad_lang=IT&ad_curr=EUR&ad_price=1.00&fallback_cids=5d9ea246b003ee05866b416e5e151b88f01a8601e477849e5d9fdba658fd40510ec7d6c8&hide_login_modal=true&share=web) and bioengineering tissue for cloning and Regenerative Medicine (RM). X-Ray Computed Tomography (CT), Magnetic Resonance Imaging (MRI) visualized with open source 3d Slicer 4.31, can be used for inspiring biomedical CADs designed with a computer interfaced with a 3D printer via the Repetier Host 0.95 D sending the G-code with the printing information (DOI: 10.1109/AQTR.2014.6857890).

Antonio Silvestro right after having being involved in the Erasmus + project in Latvia as *'Solidary Corp'* volunteer helping children of the local community teaching Yoga, music, cooking and languages classes, doubtful to keep the MSc *'Plant, Food Sciences and Environmental Biotechnology'*, decided to divulge his holistic wisdom as *Kindle Self Publisher* (KDP) on Amazon, in the *'Santosha'* of holding a Ba degree in "Biological Sciences" got with honours and awareness of the need for tools only for going ahead in the R&D. Nevertheless, between the Bachelor and the Master, he travelled in Switzerland, UK and Cyprus joining permaculture, agroforestry, aquaponics, and phytoremediation projects in countryside is developing communities, organic farms, designing environmental and wellness architectures, increasing his awareness through Yoga and Shamanism spiritual healing retreats, giving the *'HolYoga'* workshops in Denmark and Germany, both in private estates and in public festivals. Among the previous engineering designs the *'Telepika – Telescopic Pipe Moka'* of which the assembly protocol is available on Instructables and *'Aphrodite and Ares entwined in a torus inductor making the baby Hermes rototraslating!'* offered on Amazon KDP.

Recycled and Recycling Olympic 3D Bio-Printer (Recyclebot and RepRap-based)

© Copyright Antonio Silvestro, 2020

www.ingramcontent.com/pod-product-compliance
Lightning Source LLC
Chambersburg PA
CBHW040302220526
45473CB00002B/564